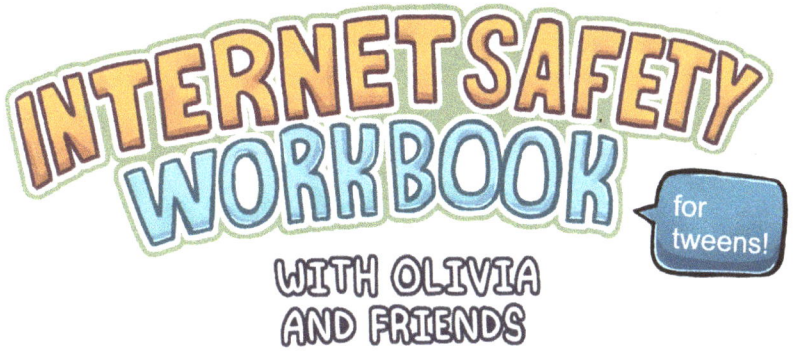

INTERNET SAFETY WORKBOOK

for tweens!

WITH OLIVIA AND FRIENDS

By **Nathan LaChine**

& Karina StarkHart, MA, LMHCA

Illustrated by Annabelle Betts

For permissions, inquiries, or additional information, please contact:
EvergreenCaregiverSupport.org

Self-published by
Evergreen Caregiver Support
Lakewood, WA, USA
ISBN: 979-8-9928534-0-7
Cover Design and Illustrations by Annabelle Betts
Co-Author Karina StarkHart, MA, LMHCA
Printed in the United States of America

Dedication

To my husband, Koon, your insights, edits, and unwavering support have been the cornerstone of this book. Without your thoughtful perspective and constant encouragement, this project would never have come to life.

The support you have given me throughout the months of writing, editing, and illustrating has been immeasurable to me. I love you more with each passing day, and I am so fortunate to share this journey and every other with the man I love.

Acknowledgements

To my nephew Michael, your youthful insights and wisdom have been invaluable. The authenticity of the dialogue and messaging in this book owe much to your eager support and patience with my endless questions throughout these months of writing.

To my brother Elliot, your insights, friendship, and guidance have been instrumental in shaping this curriculum and workbook. Our conversations, adventures, and our bond have profoundly influenced my life and this project.

And to all the kiddos we've had the privilege of caring for, our shared memories, experiences, and laughter have been the driving force behind this book.

Table of Contents

Table of Contents

About the Author

 **Nathan LaChine** is a third generation foster parent with over 20 years of experience providing therapeutic foster care to some of Washington state's most vulnerable youth. His frontline experience provides unique insights into the challenges young people face, particularly in navigating the digital world, making him a sought-after internet safety expert.

He founded Evergreen Caregiver Support as an organization dedicated to empowering caregivers through education and community resources. Through his organization Nathan has developed trainings on Commercial Sexual Exploitation of Children (CSEC), Child Sexual Abuse Material Industry (CSAM), LGBTQIA+ , SOGIE, Harm Reduction, among many other trainings.

Nathan has co-hosted a local radio show Real Family Matters, providing resources to community members. He frequently speaks to legislators and policymakers at the local, state, and national levels on issues affecting foster care, and has been featured on numerous podcasts and in interviews.

Nathan serves as a Foster Parent Mentor and Support Group Facilitator with the University of Washington, providing guidance and support to caregivers across the state. He is also a member of the Department of Children, Youth, and Families' Foster Parent 1624 Consultation Team.

Nathan's work is driven by the principles of Educate, Empower, and Harm Reduction, inspiring meaningful change through his trainings, lectures, and advocacy. He is also a published author, professional speaker, international trainer, community advocate, and avid collector of modern and historic queer art.

Find Nathan's Trainings

For more information, additional resources, full catalog of trainings, and access to the training program "Internet Safety and Warning Signs of Online Grooming" - which this book is designed to complement - visit EvergreenCaregiverSupport.org or scan the QR Code below.

About the Co-author

 **Karina StarkHart** is a licensed mental health counseling associate and sex educator with a passion for empowering youth and younger adults to navigate the complexities of today's world, including the challenges of online safety. Drawing on nearly a decade of prior experience in therapeutic foster care and social work, Karina understands the crucial role of open communication and healthy boundaries in protecting young people.

Karina currently works in private practice in Washington State, specializing in providing mental health support to adults regarding relationship, sexuality, and identity concerns, trauma recovery, and attachment issues. She has a master's in mental health counseling from Antioch University Seattle, where she graduated with honors, and holds dual bachelor's degrees in psychology and English literature from The Evergreen State College.

As a former foster parent and social worker, Karina developed and facilitated a variety of trainings for parents and professionals on topics such as trauma, boundaries, and self-care. She also facilitated a foster parent support group with her co-author, Nathan LaChine. Her experience includes working in educational settings supporting youth with high behavioral needs and volunteering with a crisis helpline.

Karina's commitment to youth well-being drives her work on internet safety, combining her expertise in mental health with practical strategies for parents and professionals to help youth navigate the digital world. When she's not counseling or writing, Karina enjoys spending time with her family and curling up on rainy days with a coffee, book, and her dog, Walter.

Connect with her at www.desireblooms.com

About the Illustrator

Annabelle Betts is a high school art kid with an ever-growing list of projects under her belt, including self-published art calendars, comics, a zine, and a coloring book. Her bucket list of future projects is equally ever-growing. She specializes in visual storytelling through both traditional and digital illustration, collage, and comics, with various mediums.

She has also dabbled in product design, patchwork, and book design, this is her first mostly-text book design project.

Check out Annabelle's comics on Tapas at @BananaBelle, her illustrations and other projects on Instagram @bananabelleart, and her art shop on Etsy, at bananabelleart.etsy.com.

Message to Adults

Before we start, I want to spend a moment to encourage you to sit with your child as they go through this workbook learning about internet safety. The internet is one of the most powerful tools humanity has ever created, and there are many great reasons for parents to encourage children to engage with online learning and online gaming. With that said, just like anything else it can be used for nefarious purposes. Unfortunately, child predators can engage with your children on a daily basis without your knowledge. They are able to communicate to your children from anywhere they have internet access, such as from their homes, their bedrooms, on the bus to school, or while they are hanging out with friends. This is why we must educate our children about online safety and the fact that sadly, not all people online are who they claim to be.

I encourage you to read this workbook with your child, one section at a time, and have fun completing the activities together! Each activity is designed to reinforce the learning objectives of the previous section. We have included several "Chat Starters" to spark great conversations about the scenarios for you to discuss together.

While this book is designed as a standalone workbook for parents and children to facilitate the discussion of safe online habits, it is also a companion to the training series "Internet Safety and Warning Signs of Online Grooming" offered by Evergreen Caregiver Support. This book builds on content presented in the training series which provides additional depth and discussion about these topics. It is my hope that with this book and the training series, you will feel empowered to have in-depth conversations with your children and improve their knowledge and understanding of safe online habits.

Message to Kids

There are lots of fun things to do online. In this book, you will get to hear the stories of several kids your age who like to play games, chat with friends, and watch videos online.

Did you know that sometimes untrustworthy people will try to use these activities to trick or hurt kids? When someone is not aware of these dangers, they might not know they are talking to a mean person until it is too late!

This might seem scary but with a little information and preparation, you can stay safe and still have all the fun you have always enjoyed online. You might even learn some helpful tips to share with friends! The following stories and activities about Alex and his friends will teach you important skills to stay safe online.

There are many ways to engage with others online.

Do you like to do any of these online activities?
What house rules are there around you doing
these activities?

-Play multiplayer games
-Chat with friends on social
media
-Make new friends
-Listen to music
-Watch TV or movies
-Watch, share, or comment on
online videos
-Create your own videos
and share them
-Download new
games and programs

_____

_____

_____

_____

_____

Internet Word Search

```
J T R O L L D U E M E M M U Y B Z C X
R Q L Z Y G Q W T S E I R A D N U O B
I X Z M O N O U K S R U D Q H C C I P
V G O G G I R V O C K H L O C G Y R T
B N D D R F X V L O G K Z C A A I G N
K I Q T O O Y W V G N O W S C V N V E
W Y O C O O I D P P A Q M Y A P P W S
L L I O M P J R J D B O N C G X E Q N
Q L D Q I S S C R D Z T Y T P T R B O
J U Y Y N C C B I J B S G G Z T S N C
F B N C G O M F X K B X P I K M O Z N
M R G N I T X E T J N P W W N U N E I
H E B X X I Z I U B R A C A O R A K K
Z B J J C U U Y K O L C Y X X X L S M
I Y F D C M P S F P E S K U K D M Y S
V C X G Z N F I K R A I L O M M D I Q
G I K I Q Z L I F T U O H S U T D E B
Z Z R Q U E K O Q E A S F M Z Q A F S
J C G U F N E R U V V T O J H U P I Z
D R O W S S A P S E I K O O C M I E O
```

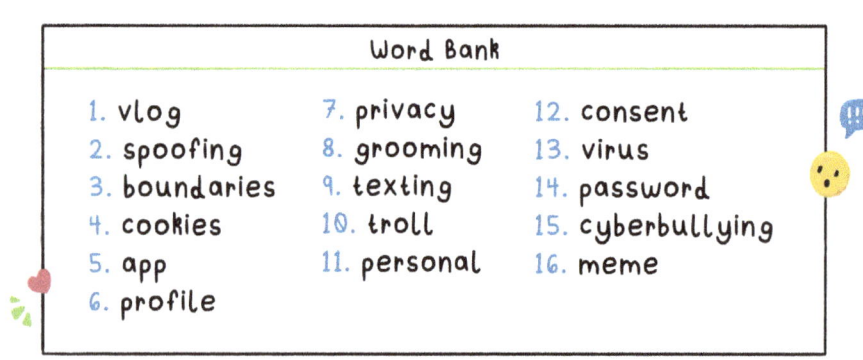

Word Bank

1. vlog
2. spoofing
3. boundaries
4. cookies
5. app
6. profile
7. privacy
8. grooming
9. texting
10. troll
11. personal
12. consent
13. virus
14. password
15. cyberbullying
16. meme

(Answer key on page 71)

Now it's time to check in on our friends and see just how safe they are being online.

Olivia Gets Tricked

Meet Olivia! She's 12 years old and loves online gaming, which is where you can find her after school and on weekends. Whether she's mastering new levels, exploring virtual worlds, or teaming up with friends online, Olivia always brings her creativity and quick thinking to the game. When she's not gaming, she's learning new tricks and sharing tips with her gamer friends.

Olivia recently made a new friend named John from school, who goes by the gamer tag "JohntheConqueror". Olivia recently accepted a friend request from him on social media. This seemed odd because John had told her that he did not have a social media page.

John and Olivia text each other a bit through social media. And then, John sends Olivia a link through his Messenger. It says: "Click here for Top 5 Cheats!"

Olivia thinks that John is sharing some cheat codes for the game with her.

Olivia clicks on the link, but nothing happens. She thinks maybe the link is broken, so she leaves it and doesn't think about it anymore. Since the next quest isn't until tomorrow, she decides to go to bed and call it a night.

What could have happened when Olivia clicked on the link? Have you ever clicked on a random link you were unsure of?_____

_____

_____

_____

The next morning, Olivia tries to check her social media accounts, but she's logged out. She gets really frustrated as she tries to log back in, but none of her passwords are working. No matter what she tries, she can't get into her accounts.

After getting to school, Olivia goes to the normal meet-up spot with her friends. They all ask her about all the messages and links she sent out last night.

Olivia is confused as she has no idea what they're talking about, but she's positive that she didn't send any links or messages to her friends last night.

Olivia wonders what is going on. And then, she just so happens to run into John in the hallway.

"What were the cheat codes you sent me yesterday?" Olivia asks. "I clicked on the link, but nothing happened and now I can't log into my accounts!"

John is confused.

"What are you talking about?" he asks Olivia. "You know my parents won't let me have a Social Media account!"

What do you think happened to Olivia?

_____

_____

_____

_____

What would you have done if you were her?

_____

_____

_____

_____

B _ _ n d _ r _ e _ : The rules or limits we set to keep ourselves and others safe, happy, and respectful.

C _ _ s _ _ t: Means saying "yes" or agreeing to something, but only if you really want to.

C _ _ e r b _ l _ y _ n g: Bullying that happens online through mean messages, posts, or sharing hurtful things.

E _ _ j _ : Small icons like 😊 or 😎 used in messages to show feelings or reactions.

M _ _ r o t _ _ n s _ c t _ _ _ n s: Are small purchases made within a video game, usually with real money, to buy virtual items or unlock special features.

P _ _ s _ _ r d: A secret code you create to keep your accounts safe from others.

P _ _ s _ n a l I _ _ o _ _ a t i _ _ _ : Details like your full name, address, phone number, or school that you shouldn't share online.

P _ _ s _ i n g: A trick where someone pretends to be a trustworthy person to get your personal info, like passwords.

P _ _ v _ c y S _ _ t _ _ g s: Tools on apps and websites that let you control who can see your information.

S _ _ m: A deceptive scheme intended to trick people into giving away money or personal information.

S _ x _ _ r t i _ _ : When an online predator tricks some -one into giving them nude images or videos, and then demands money, more images, or makes other demands.

S _ _ m: Unwanted or junk emails or messages, often from strangers or companies.

S _ _ o _ i n g: Is when someone pretends to be someone or something they're not, usually to trick others.

T _ _ l _ : Someone who says mean or annoying things online just to upset others.

T _ _ - f _ _ t _ r a _ _ h _ _ t _ f i c _ _ _ i o n: A security feature that requires two forms of identification (like a pass -word and a code sent to your phone) to log into an account.

(Answer key on page 72)

Angel and an Online Friend

Let's check in on Olivia's friend, Angel, who is always on the move, either playing sports (soccer or baseball), hanging out with friends, reading, and playing video games. Angel just got their own bedroom for the first time and has spent the ENTIRE weekend setting up their new room.

Angel is proud of how their room turned out and wants to show off how cool it is to Olivia, Alex, Liam and some of their online friends.

What do you see?

AL.angel_x

2 minutes ago

Circle as many of Angel's interests and hobbies as you can find in this picture that was sent.

Does Angel's room look like yours? Do you and Angel have anything in common? What items in your room tell a story about who you are and what you like? How about where you live? _____

_____

_____

_____

Angel decides to share a picture of their decked-out room with their friends and their friends' families on social media. Angel is excited to show off all their brand-new soccer gear, all their trophies, and of course, their computer gaming setup.

Angel posts the picture online with the caption, "Welcome to my world!"

Do you know how Angel can safely share this picture where only their friends and family can see it?

_____

_____

Parents: Walk through these steps with your child.

There are three options available for Angel when sharing their photo online.

What is the difference between:

Share with everyone: _____
_____

Share with Friends/Family: _____
_____

Share with Public: _____
_____

Can private pictures ever get out to people you didn't send them to (get "leaked")? How? _____
_____
_____
_____

Angel doesn't pay any attention to these options and just posts the pic publicly. Soon, Angel's friends start commenting on the photo. They say nice things about the trophies and especially like the gaming setup. Some parents even comment about how tidy the room looks!

Angel also gets a comment from a user called "Joey14forever", Angel doesn't recognize who it is...

How do you think Angel is feeling getting all these comments? How would you feel?

_____

_____

_____

Before long, "Joey14forever" starts messaging Angel. This person says that their name is "Joey," but that Angel could call him "Joe." Joe says that he thinks they go to the same school as Angel, and Joe wants to know the name of Angel's school to see if he is right.

How would you feel if someone you didn't know started commenting and chatting with you?

_____

_____

What concerns should Angel have about this person?

_____

_____

_____

Soon, Angel and Joe are talking every day about sports, video games, school, and whether they have a crush on anyone at school. Joe tells Angel they are "really mature" and "nice" to talk to.

How does it feel when someone compliments you?

_____

_____

How do you feel when someone notices you sound grown up?_____

_____

_____

After a couple of weeks, Angel realizes that they have shared a lot about themself with Joe. But they know so little about him! Angel starts to wonder why the two of them have never met up at school even though they talk online every day. Angel decides to ask Joe about it.

When Angel asks Joe why they never met up at school, Joe says that he "actually goes to a different school nearby.".

Should Angel be concerned? What should Angel ask next?_____

_____

_____

A few days later, Joe tells Angel that they should meet up at the park near Angel's school on the weekend and hang out.

Should Angel meet up with Joe? _____

_____

_____

What would you do? _____

_____

_____

Who would you talk to about this suggestion before agreeing or disagreeing? _____

_____

_____

Before meeting up, can you answer yes to all these questions about your online friend?

☐ Have you met them before?
☐ Do you know them from school?
☐ Do you know what grade they are in?
☐ Have your parents met them before?
☐ Have you met their parent's before?
☐ Have your parents met their parents?
☐ Do you know their last name?
☐ Do you know their address and phone number?

Your House	Mall
Their House	Local Park
Another Friend's House	Your Treehouse
Library	Where Else?

What makes a place safer or less safe? _____

_____

_____

Angel thinks about it for a long time. In the end, they decide to be smart and do the right thing. Angel talks to their parents about Joe and says that they are uncomfortable about meeting him at the park.

- GOOD JOB ANGEL! -

Angel's parents are glad that they came to them and told them about it before making a decision. They tell Angel, "While chatting and meeting new people online can be fun, meeting up with people that you first met online can be dangerous. People online can misrepresent themselves and you always should talk with us or a trusted adult before meeting new people for the first time."

Angel's parents agree to go with them to the park if they still want to meet their friend Joe. They ask for Joe's parents' phone numbers to call ahead and have them come too. Now Angel can be safe and have fun!

What do you think would happen if Joe was not who they said they were? _____

_____

_____

Angel and their parents talk about what information, pictures, and videos are okay to share online and what should be kept private.

What would you suggest Angel not share online?

_____

_____

_____

Together, Angel and their parents create their very own Internet Usage Agreement to make sure Angel knows what they can and cannot do on the Internet.

What would you put into this agreement? See an example on page 63._____

_____

_____

_____

_____

Internet Term Crossword!

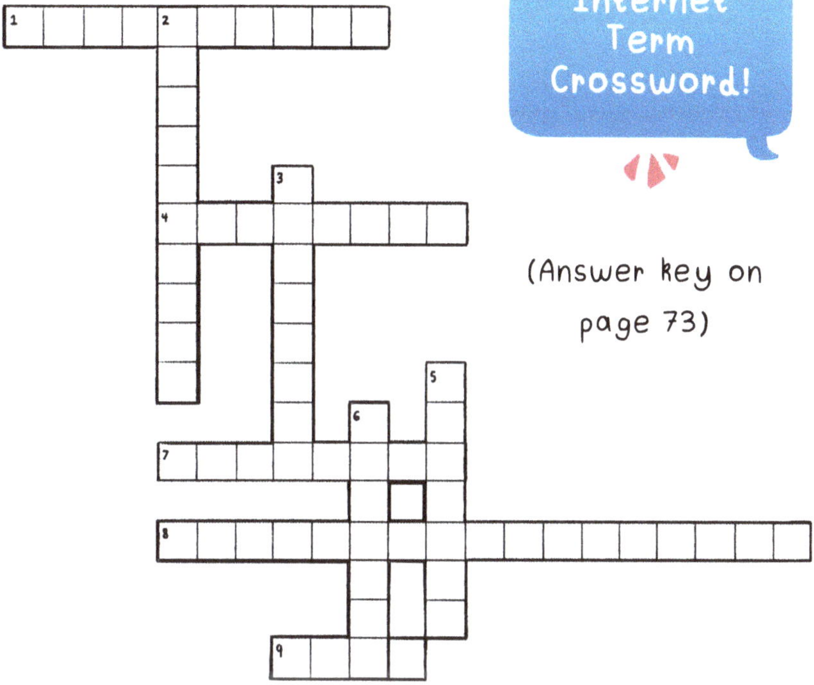

(Answer key on page 73)

Across

1. When content on screen is saved as a picture.

4. A trick where someone pretends to be a trustworthy person to get your personal info, like passwords.

7. Software that helps protect your device from viruses and hackers.

8. Small purchases made within a game or other app, usually with real money, to buy virtual items or unlock special features.

9. Unwanted or junk emails or messages, often from strangers or companies.

Down

2. A process of turning data into a secret code to protect it, ensuring only the intended recipient can read or use it.

3. A name you choose to represent yourself online, like a nickname.

5. Harmful software that can damage your device or steal your information.

6. A security test on websites to make sure you're a real person, not a bot. It usually asks you to type characters or click on certain images.

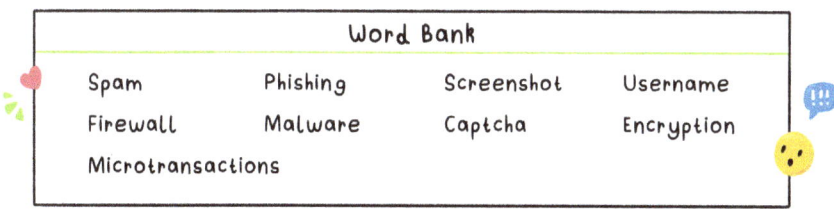

Word Bank

Spam	Phishing	Screenshot	Username
Firewall	Malware	Captcha	Encryption
Microtransactions			

Alex and the Picture

This is Alex!

Alex is 12 years old and he loves spending time on his phone. Ever since his parents got him his own phone, he has been spending more and more time on social media.

This is Maggie, she is 13 years old, and is Alex's crush from school. Alex adores her and loves it when she pays attention to him. Maggie can be a little annoying sometimes, but Alex doesn't mind.

Alex and Maggie have been messaging each other for a few months. They share a lot of common interests and get along fabulously.

Even the lock screens of their phones are pictures of each other. They text and send snaps to each other daily, it has become one of their habits as soon as they wake up. Right now, they are on an 80-day snap streak.

Children: How long do you talk to someone before you tell your parents you have a crush on them?

_____

_____

Parents:
How long before they "make it official" would you want to know? Have you talked with your pre-teens about relationships and dating?

TALKING TIPS FOR PARENTS:

Did you know that pediatricians suggest kids wait until they are at least 16 years old to start dating but kids as young as 10 or 11 years old might start exploring dating relationships on their own? Waiting to talk to your kids until you think they are ready can sometimes lead to missed opportunities to prepare them ahead of time to make healthy choices. Take a moment now to have an open conversation with your child. Ask questions with curiosity and work towards making agreements that your child feels invested in, rather than setting "rules" they might not understand the reasons for.

CHAT STARTERS:

What does it mean to have a "girlfriend" or boyfriend" at different ages?

What does flirting mean to you? What does it look like?

What do you think most people are excited about when they choose to start dating?

How do you think you will know you're ready to start dating?

Who will you talk to and seek advice from about your dating life?

What examples have you seen of really successful couples? What made them successful?

One day something unexpected happened. Maggie asks Alex for shirtless pictures. Maggie says that she won't show the pictures to anyone. Alex feels unsure but he wants to make Maggie happy.

How do you know when something is not okay with you? What does that feel like? Have you ever told a friend "No" before? How did it go? What other things should Alex consider first? _____

_____

_____

_____

_____

Alex doesn't want to disappoint Maggie, so he sends a pic of himself without a shirt on. He tells himself it's fine to send just one picture. Maggie tells him how much she loves his picture. Alex feels good that Maggie likes it, but Alex still feels unsure if it was a good idea. Pretty soon, Maggie asks him for more pictures. Alex thinks about it for a moment before he decides that it would be okay since he already sent one pic.

Can you think of any reason sending a picture of yourself (or someone else) without all of your clothes on might be a bad idea? Even just once? _____

_____

_____

_____

A couple of days later, Maggie tells Alex she wants to break up. Alex is really upset because he didn't see it coming. He thought Maggie was the one, and now she won't even reply to his texts. He thought things were getting better between them!

Alex decides to ask his friends if they've talked to Maggie since the breakup, but they say they haven't. Then they tell him that Maggie sent the pictures of him without his shirt on to all her friends and some of her friends started showing them to everyone! Soon, Alex's pictures had been shared with everyone at school. Someone even printed them out and put them in the bathrooms and on the bulletin board.

Alex can't believe what's happening. Everyone at school is laughing at him and making fun of him. He feels like he's going to need help from his parents and the school counselor because he doesn't know what to do. He's even thinking he might need to tell the police.

Chat Starter: Have you ever been in a situation like this or known anyone who has? What happened and what could have been done to make it better?

What could Alex have done differently to avoid this situation? Do you know anyone who something like this has happened to? _____

_____

_____

_____

_____

Tip: Your parent will always be a great support, even when things go wrong. Did you know that temporary or expiring pics can be screenshotted and saved?

😎	☐	☐ Can't say
👈👉	☐	☐ Foolish
👀	☐	☐ Cool or spicy
👁👄👁	☐	☐ Cool
🙃	☐	☐ Sarcastic
🫠	☐	☐ Shy or nervous
🤢	☐	☐ Stunned
🤡	☐	☐ Disgust
🤫👂	☐	☐ Mind Blown
💀	☐	☐ Look or curious
💅	☐	☐ Slay queen
🔥	☐	☐ No way or embarrassing
🙏	☐	☐ Look or curious
🤫	☐	☐ Keep it secret

Many people use emojis in their own way, if you are unsure what someone means, don't be afraid to ask.

(Answer key on page 73)

Liam's Adventures in Online Gaming

Liam just met a new gamer online, and now he's wondering if they've found a new teammate, a special someone, or... something else? Liam has been climbing the ranks in a bunch of his games and spends hours every day playing on his computer. He is always chatting with his teammates to set up quests, tournaments, and share stuff in the game. One of his teammates, who goes by the name "BlondBombShell," has been talking with him for months, and they've become two of the top players in their guild.

"BlondBombShell" says that her name is Jessica and that she is also 12 years old like Liam. They are always coming up with new strategies together to stay ahead of the other players. Jessica says it would be more efficient if they talked off the game using another chat program. Liam agrees as he is always looking for a competitive edge.

Has anyone ever asked to chat with them outside of the game you're playing?_____
_____
_____

Liam and Jessica continue to dominate the game and spend more and more time off-site using the chat program she suggested. Jessica asks Liam lots of questions and even exchanges pictures with him. One night, while they're on a quest in a game, Jessica says, "You know, you're kinda cute!" Jessica asks Liam if he could send her a couple pics without his shirt on.

What should Liam do? _____
_____
_____

Has anyone ever asked you for a picture or video of yourself that felt very private or made you feel uncomfortable? _____
_____
_____

Liam sends a couple shirtless pics to Jessica because he also thinks that she's hot. Jessica responds to his photos with, "Hottie, I want to see more!"

Liam thinks about whether he should send her more photos. He is fond of Jessica after all. Jessica says that she will send him some more photos of herself if he also sends some more.

What should Liam do? Should he send more pictures, or should he stop? _____

_____

_____

Jessica continues to ask for more pictures of Liam with less and less clothing, Liam starts to feel uncomfortable because she keeps asking for more pictures. And then one day, she asks Liam for a picture of him without any clothes on. Liam firmly says, "No!"

But Jessica does not accept his answer. "If you don't send me more pictures, I am going to post the pictures you have sent me to your school's social media page, soccer team's page, and to your friends and family!" she threatens him.

Parent tip: Have you ever shared with your child about a mistake you made when you were young? While very young children need their parents to be strong and endlessly capable, older children can benefit from hearing that you are a person too and that everyone struggles with something at some point.

Liam doesn't want to get into trouble. He hopes that if he sends Jessica a picture, she will leave him alone. So, he does as he is told...

Unfortunately, Jessica does not stop bothering Liam, even after that. She demands more pictures and even a video! Liam, not knowing what to do, decides to block Jessica on all of his social media and game accounts and hopes that no one will find out.

Is there anything else Liam could have done besides blocking Jessica?_____
_____
_____

What would you do?_____
_____
_____

A few days later, Liam receives a random message from an account he doesn't recognize. The message he receives includes a screenshot of his photos pending posting to his soccer team's social media page. The message is from Jessica asking, "why did you block me?" Liam starts freaking out and doesn't know what to do.

What could Liam do at this moment?_____
_____
_____

What would you do? Who could you talk to about this?_____
_____
_____

Liam doesn't know what else to do and he's afraid that other people will find out what he has done. He cannot imagine what would happen if any of his friends saw the photos he sent to Jessica. So, he sends Jessica more pictures and a video hoping that she goes away and finally leaves him alone.

Can you help Liam out? What would you tell him to do?_____
_____
_____

Even after Liam sends more pictures and videos, Jessica doesn't go away. She keeps bothering him and he becomes more and more stressed with each passing day. After several weeks, Liam finally decides to own up to his mistakes and tell his parents about everything that has happened.

Liam's parents are glad that he told them about it, and say he should have told them about his situation before things got out of hand. They say that what Jessica is doing is called "Sextortion" and that it's against the law. They assure Liam that he will not get into any trouble and Jessica can be stopped with their help. Liam and his parents reached out to the local law enforcement to report Jessica.

Remember that if you are in a similar situation, you can tell a trusted adult instead. There are solutions, but agreeing to never share pictures in the first place can serve as prevention.

Did you put into your Internet Usage Agreement that you would not share these types of pictures?

_____

_____

What would you do if this happened to you or one of your friends? _____

_____

Do you think Jessica was who she said she was? What helps you to know? _____

_____

_____

A few weeks have passed since Liam told his parents and the police about Jessica. Since then Jessica has stopped bothering him and he is not getting any more random messages from unknown accounts. Now, Liam wants to make sure his friends don't go through the same thing he did, so he decides to give them some advice about chatting online.

What should Liam tell his friends, so they don't get tricked into this situation?_____

_____

_____

_____

_____

We spend more and more time on the Internet every day. Just like what happened to Olivia and her friends we can also get caught up in dangerous situations if we're not careful.

What have you been told about how to stay safe online?_____

_____

What advice would you give Olivia or any of your friends if they told you that this was happening to them?_____

_____

_____

_____

PARENTS TIPS:

It is normal to feel afraid for your child, and often parents are tempted to be stern or impose discipline in situations like this as a way to feel safer and more in control. However, children faced with the above scenarios are "victims" and more than anything they need your love and support. When in doubt, bring kids in closer by telling them you love them and that you "have their back." This creates and maintains strong trust bonds that help kids feel confident to come to you with their concerns in the future, possibly even before trouble begins. By leaning into love and education, your kids will have the greatest possible chance of success in difficult situations, now and in the future.

Internet Safety and Usage Contract Example

Never Share Personal Information for example:
- Name
- Birthday/Age
- Address
- Phone number
- Email address
- School
- City/state you live in

Never Talk to people online that you do not know from school, family, or neighborhood

Never click on unknown links or download random files

Never post or send mean messages

Never send pictures or videos to people you don't know

Never keep secrets about something that makes you uncomfortable

Never share your passwords (except with parents/guardians)

I will not open a new account or change passwords without my parents/guardians knowledge

I will only message the below people that my parent/guardian has approved:

People I can message online (Real name)	Online username

_____   _____
Child Signature   Parent/Guardian Signature

Internet Safety and Usage Contract Example

I won't share personal information online

I won't post or send inappropriate photos or videos

I won't talk to strangers online or meet them in person

I won't share my passwords
(except with my parents/guardians)

I won't Click on Suspicious Links or Download
Unknown Files

I won't overshare on social media

I won't Cyberbully or engage in online drama

I promise never to trust everything I see online

I'll never assume what I post online is private

I'll never ignore Red Flags/Gut Feelings - I will tell an
adult if something feels wrong

- _____
- _____
- _____
- _____
- _____

_____   _____
Child Signature   Parent/Guardian Signature

Now it's time for you and your parents / guardians to write your very own Internet Usage Agreement.

- _____
- _____
- _____
- _____
- _____
- _____
- _____
- _____
- _____
- _____
- _____
- _____
- _____
- _____
- _____

_____   _____
Child Signature   Parent/Guardian Signature

Final Thoughts

Pre-teens and teens will start facing all kinds of challenges as they grow and will need you more than ever as they learn about the complexities of the world. How do you initiate conversations with your child? What activities do you do together that help create opportunities for discussion? Ask your child now what helps them feel confident to open up to you and what prevents or discourages them from sharing. Be open to their responses - they are giving you a roadmap for their needs.

Educate them about their digital world and the potential dangers that exist online. **Empower** them to make informed, safe, and smart decisions about who they interact with online, what information they share, and the potential risks of sharing images and videos. **Harm Reduction** is achieved by educating them about the internet, showing them how to report issues, and reminding them that you ALWAYS have their back!

Parents need support too when difficult situations come up. Who do you go to when you need to talk out challenging situations? If you haven't already, consider making a list or adding certain contacts as "favorites" on your phone to remind you that you also have support when needed.

Remember that you are always in control of your internet usage. If you run into problems, there is help for you 24/7, never let the perpetrator steal your voice. Use your voice and report them to a trusted adult, you deserve to be safe online!

☆ OLIVIA ♡

Answer Keys!

J T R O L L D U E M E M M U Y B Z C X
R Q L Z Y G Q W T S E I R A D N U O B
I X Z M O N O U K S R U D Q H C C I P
V G O G G I R V O C K H L O C G Y R T
B N D D R F X V L O G K Z C A A I G N
K I Q T O O Y W V G N O W S C V N V E
W Y O C O O I D P P A Q M Y A P P W S
L L I O M P J R J D B O N C G X E Q N
Q L D Q I S S C R D Z T Y T P T R B O
J U Y Y N C C B I J B S G G Z T S N C
F B N C G O M F X K B X P I K M O Z N
M R G N I T X E T J N P W W N U N E I
H E B X X I Z I U B R A C A O R A K K
Z B J J C U U Y K O L C Y X X X L S M
I Y F D C M P S F P E S K U K D M Y S
V C X G Z N F I K R A I L O M M D I Q
G I K I Q Z L I F T U O H S U T D E B
Z Z R Q U E K O Q E A S F M Z Q A F S
J C G U F N E R U V V T O J H U P I Z
D R O W S S A P S E I K O O C M I E O

How did you do?!

71

Fill in the blanks

Answer Key

Boundaries: The rules or limits we set to keep ourselves and others safe, happy, and respectful.

Consent: Means saying "yes" or agreeing to something, but only if you really want to.

Cyberbullying: Bullying that happens online through mean messages, posts, or sharing hurtful things.

Emoji: Small icons like 🙂 or 😎 used in messages to show feelings or reactions.

Microtransactions: Are small purchases made within a video game, usually with real money, to buy virtual items or unlock special features.

Password: A secret code you create to keep your accounts safe from others.

Personal Information: Details like your full name, address, phone number, or school that you shouldn't share online.

Phishing: A trick where someone pretends to be a trustworthy person to get your personal info, like passwords.

Privacy Settings: Tools on apps and websites that let you control who can see your information.

Scam: A deceptive scheme intended to trick people into giving away money or personal information.

Sextortion: When an online predator tricks some -one into giving them nude images or videos, and then demands money, more images, or makes other demands.

Scam: Unwanted or junk emails or messages, often from strangers or companies.

Spoofing: Is when someone pretends to be someone or something they're not, usually to trick others.

Troll: Someone who says mean or annoying things online just to upset others.

Two-factor authentication: A security feature that requires two forms of identification (like a pass -word and a code sent to your phone) to log into an account.

Glossary

Ad Blocker: A software tool or browser extension that prevents ads from appearing on websites to improve the browsing experience.

Antivirus: A software program designed to detect, prevent, and remove harmful computer viruses and other malicious programs.

App: Short for "application" it is a software program designed to perform specific tasks on devices like smartphones, tablets, or computers.

Binge-watching: Watching multiple episodes of a TV show or series in one sitting, commonly done on streaming platforms.

Blog: A website where a person or group regularly writes posts, typically about a specific topic or personal experiences.

Blogger: A person who regularly writes and shares material, usually on their blog.

Boundaries: The rules or limits we set to keep ourselves and others safe, happy, and respectful.

Captcha: A security test on websites to make sure you're a real person, not a bot. It usually asks you to type characters or click on certain images.

Chatroom: An online space where multiple people can communicate with each other in real-time through text.

Cloud Storage: A service that allows you to save files on the internet instead of your computer, making them accessible from any device with internet access.

Consent: Means saying "yes" or agreeing to something, but only if you really want to.

Cookies: Small files that websites save on your computer to remember your preferences, login details, or track your activity.

Cyberbullying: Bullying that happens online through mean messages, posts, or sharing hurtful things.

Download: The process of transferring data from the internet to your device, such as files, apps, or games.

Glossary

Digital currency: A type of money that exists only in electronic form and is used for online transactions.

Direct Message (DM): A private message sent directly to someone online.

Emoji: Small icons used in messages to show feelings or reactions.

Encryption: A process of turning data into a secret code to protect it, ensuring only the intended recipient can read or use it.

Fake News: Made-up stories online that aren't true. Always double-check before believing or sharing.

Firewall: Software that helps protect your computer or device from viruses or hackers.

Forum: An online discussion board where people can post questions and share answers about various topics.

Ghosting: The act of suddenly cutting off all communication with someone without explanation, often seen in friendships or dating.

Grooming: When someone online tries to build a friendship with you in order to take advantage of you later.

Malware: Harmful software that can damage your device or steal your information.

Meme: A humorous image, video, or text that is copied and spread rapidly online, often with slight variations.

Microtransactions (in-game): Are small purchases made within a video game, usually with real money, to buy virtual items or unlock special features.

NSFW: Stands for "Not Safe For Work." It indicates that content is inappropriate for a workplace or school environment, usually due to adult themes or language.

Open Source: Software that is freely available for anyone to use, modify, and distribute, often developed collaboratively by a community.

Password: A secret code you create to keep your accounts safe from others.

Glossary

Personal: Means something that belongs to or relates specifically to you, such as your feelings, belongings, or choices.

Personal Information: Details like your full name, address, phone number, or school that you shouldn't share online.

Phishing: A trick where someone pretends to be a trustworthy person to get your personal info, like passwords.

Podcast: Podcasts are discussions published online. They are a bit like radio programs.

Privacy: The right to keep your personal information, actions, and thoughts hidden from others unless you choose to share them.

Privacy Settings: Tools on apps and websites that let you control who can see your information.

Profile: Information about you that everybody can see, that can include your hobbies, favorite sport, music, etc.

QR code: A type of barcode that, when scanned, takes you to a website or shows specific information.

Registering: Signing up to use a new service, like a game, website, or chat program, by providing details like your username and password.

Scam: A deceptive scheme intended to trick people into giving away money or personal information.

Screenshot: When you save the content on the screen as a picture.

Search Term: A word that you type in when you search for information about something online.

Sexting: The action of sending sexually explicit photographs, videos or messages.

Sextortion: When an online predator tricks someone into giving them nude images or videos, and then demands money, more images, or makes other demands.

Social media: Activity that takes place on the internet on specific sites designed to connect people.

Glossary

Social media Influencer: A well-known person who is followed on social media.

Spam: Unwanted or junk emails or messages, often from strangers or companies.

Spoofing: Is when someone pretends to be someone or something they're not, usually to trick others.

Streaming: Watching videos or listening to music online without downloading the files.

Texting: Sending short written messages using a phone or other device to communicate quickly with others.

Troll: Someone who says mean or annoying things online just to upset others.

Two-factor authentication (2FA): A security feature that requires two forms of identification (like a password and a code sent to your phone) to log into an account.

Username: A name you choose to represent yourself online, like a nickname.

Virus: A harmful software program that spreads between computers, often without the user's knowledge, and can damage files, steal data, or disrupt how the computer works.

Vlog: Website or social media account where a person regularly posts short videos.

Vlogger: Person who regularly posts short videos to a vlog.

Virtual Private Network (VPN): A tool that helps protect your privacy online by hiding your real location and encrypting your internet activity.

Website: An internet page containing text, sound, images or videos.

Resources

Evergreen Caregiver Support

evergreencaregiversupport.org

ECPAT

ecpat.org/bill-of-rights

Interland

beinternetawesome.withgoogle.com/en-us/interland

Interpol

interpol.int/en/Crimes/Crimes-against-children

National Center for Missing and Exploited Youth (NCMEC)

missingkids.org

NetSmartz

netsmartzkids.org

Take it Down

takeitdown.ncmec.org

Thorn

thorn.org

> **Crisis Text Line**
> text "HOME" to 741741
>
> **The Trevor Project**
> thetrevorproject.org
>
> **988 Lifeline**
> 988lifeline.org

Reviews

"An amazingly effective tool for parents and tweens to navigate the exciting but also perilous world of social media. Written and illustrated in ways that appear to engage tweens while also offering on-the-go guidance to parents."

-Murray David Schane, M.D. President, MaleSurvivor

"As a direct service provider operating the first, and only, safe house in the country for male youth who have been sex trafficked, I have seen firsthand the alarming increase in
online exploitation, particularly among boys on gaming platforms and social media. This workbook is a great resource for helping youth recognize online threats and their own vulnerabilities in a developmentally appropriate way. It also provides excellent language for parents, teachers, and caregivers to use when speaking to the kids in their life."

-Landon C. Dickeson, MS, LPC, C-DBT, Chief Operating Officer for Ranch Hands Rescue and Bob's House of Hope

"An essential guide for parents, caregivers, and educators looking to navigate the complexities of online safety with young audiences. This thoughtfully crafted resource goes beyond simply outlining risks: it fosters meaningful discussions about digital awareness, online etiquette, and self-protection in an increasingly connected world."

-Ena Lucia Mariaca Pacheco, Human Security Expert and Researcher in Child Abuse, Exploitation, and Trafficking

www.ingramcontent.com/pod-product-compliance
Lightning Source LLC
Chambersburg PA
CBHW051550120626
46551CB00013B/1449